Machines and Motion
Screws
by Kirsten Chang

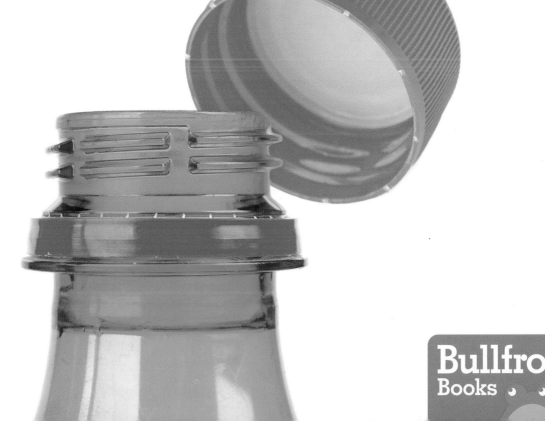

Bullfrog Books

Ideas for Parents and Teachers

Bullfrog Books let children practice reading informational text at the earliest reading levels. Repetition, familiar words, and photo labels support early readers.

Before Reading

- Discuss the cover photo. What does it tell them?

- Look at the picture glossary together. Read and discuss the words.

Read the Book

- "Walk" through the book and look at the photos. Let the child ask questions. Point out the photo labels.

- Read the book to the child, or have him or her read independently.

After Reading

- Prompt the child to think more. Ask: Screws are all around. Where do you see them? What do they do?

Bullfrog Books are published by Jump!
5357 Penn Avenue South
Minneapolis, MN 55419
www.jumplibrary.com

Library of Congress Cataloging-in-Publication Data

Names: Chang, Kirsten, author.
Title: Screws / by Kirsten Chang.
Description: Minneapolis, MN : Jump!, Inc., [2018]
Series: Machines and motion
"Bullfrog Books are published by Jump!"
Audience: Ages 5–8. |Audience: K to grade 3.
Identifiers: LCCN 2017055731 (print)
LCCN 2017056399 (ebook)
ISBN 9781624968594 (e-book)
ISBN 9781624968570 (hardcover : alk. paper)
ISBN 9781624968587 (pbk.)
Subjects: LCSH: Screws—Juvenile literature.
Simple machines—Juvenile literature.
Classification: LCC TJ1338 (ebook)
LCC TJ1338 .C445 2018 (print) | DDC 621.8/82—dc23
LC record available at https://lccn.loc.gov/2017055731

Editor: Kristine Spanier
Book Designer: Molly Ballanger

Photo Credits: Art65395/Shutterstock, cover; mihalec/Shutterstock, 1 (cap); anaken2012/Shutterstock, 1 (bottle); Africa Studio/Shutterstock, 3; ZouZou/Shutterstock, 4 (foreground); Skylines/Shutterstock, 4 (background); Grigvovan/Shutterstock, 5, 23br; Jaruek Chairak/Shutterstock, 6–7, 23tr; Oleksandrum/Shutterstock, 8–9, 23bl; Sabir Babayev/Shutterstock, 9; Andrey _ Popov/Shutterstock, 10–11; leungchopan/Shutterstock, 12; Marek Walica/Shutterstock, 13 (chair); Luis Molinero/Shutterstock, 13 (child); AVAVA/Shutterstock, 14–15 (man); Elena Schweitzer/Shutterstock, 14–15 (art); zlikovec/Shutterstock, 15; BennTheBeats/Shutterstock, 16–17; 23tl; jxfzsy/iStock, 18; Jiri Hera/Shutterstock, 19; Juice Images/Alamy, 20–21; Stocksnapper/Shutterstock, 24.

Printed in the United States of America at Corporate Graphics in North Mankato, Minnesota.

Table of Contents

What is a screw?

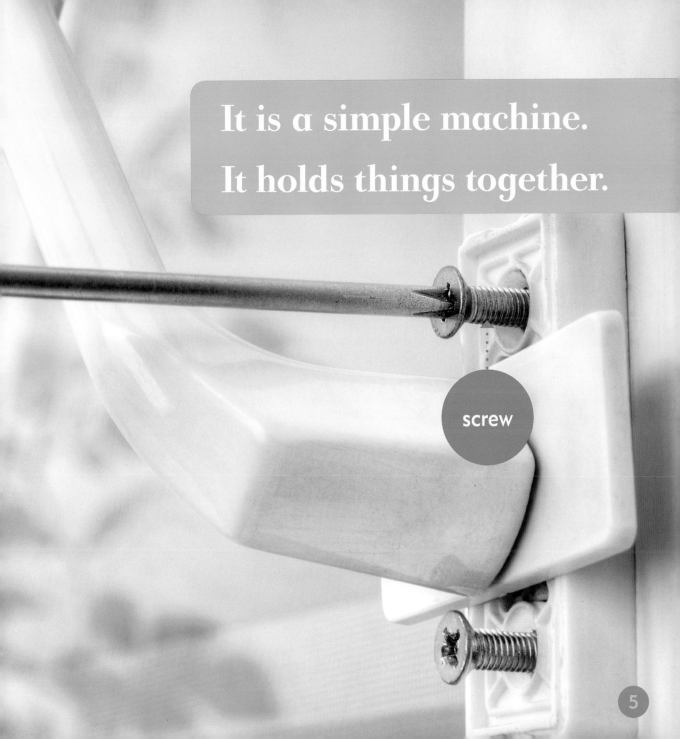

It is a simple machine.
It holds things together.

screw

A screw is a rod.

A spiral wraps around it.

rod

spiral

Some screws end in a point.
This helps cut.

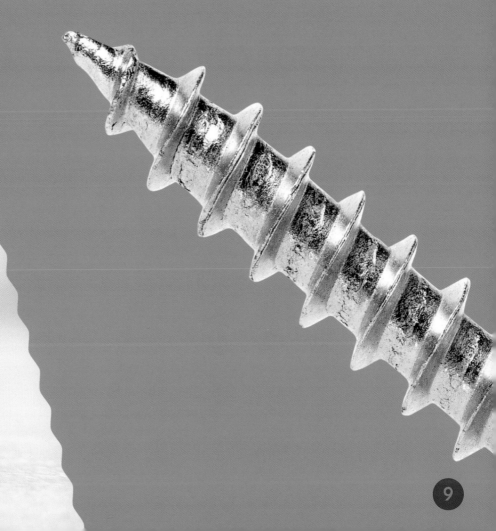

The screw turns.

Then it stays
in place.

Mom builds a chair.
She uses screws.

Wow! It is strong.

Dad hangs a frame.

He uses a screw.

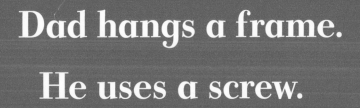

A bolt is a type of screw.

It has a flat end.

A nut keeps a bolt in place.

Bolts hold things together.

nut

bolt

What else is a screw?
Part of a light bulb.

screw ····▶

A jar lid.

screw

How do you use screws?

Use a Screw

Have an adult help you with this simple activity that compares screws and nails.

You will need:

- screw
- nail
- piece of cardboard

Directions:

❶ Push the nail through the piece of cardboard.
❷ Twist the screw through the cardboard.
❸ Try to pull the nail out.
❹ Try to pull the screw out.
❺ What did you notice? Which one was easier to pull out of the cardboard? What does this tell you about how screws hold things together?

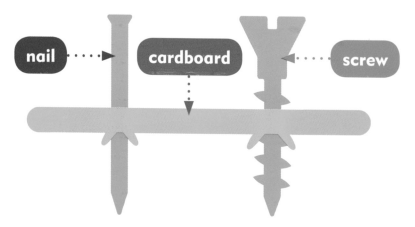

Picture Glossary

nut
A piece of metal with a hole in the middle that fastens to a bolt.

simple machine
A tool used to make work easier, such as an inclined plane, lever, pulley, screw, wedge, or wheel and axle.

screw
A spirally grooved cylinder that usually has another grooved hollow cylinder into which it fits.

spiral
A line that curves around a center.

Index

To Learn More

Learning more is as easy as 1, 2, 3.

1) Go to www.factsurfer.com

2) Enter "screws" into the search box.

3) Click the "Surf" button to see a list of websites.

With factsurfer.com, finding more information is just a click away.